SCIENCE ON THE INTERNET

A STUDENT'S GUIDE
2000 - 2001

ANDREW T. STULL
HARRY NICKLA

PRENTICE HALL
Upper Saddle River, NJ 07458

Editorial Director: *Paul Corey*
Editor in Chief Science: *Carol Trueheart*
A.V.P. Production & Manufacturing: *Dave Riccardi*
Special Projects Manager: *Barbara A. Murray*
Manufacturing Manager: *Trudy Pisciotti*
Formatting: *Jonathan Boylan*
Supplement Cover Manager/Designer/Illustrator: *Paul Gourhan*

© 2000 by **PRENTICE-HALL, INC.**
Pearson Education
Upper Saddle River, NJ 07458

TRADEMARK INFORMATION:

Microsoft Windows and Microsoft Internet Explorer are trademarks of Microsoft Corporation; Macintosh is a trademark of Apple Corporation; NCSA Mosaic is a trademark of the National Center for Supercomputing Applications; Netscape is a trademark of Netscape Communications Corporation; Java is a trademark of Sun Microsystems. All other products and trademarks are the property of their respective owners.

Printed in the United States of America

10 9 8 7 6 5 4 3 2 1

ISBN 0-13-028253-7

Prentice-Hall International (UK) Limited, *London*
Prentice-Hall of Australia Pty. Limited, *Sydney*
Prentice-Hall Canada, Inc., *Toronto*
Prentice-Hall Hispanoamericana, S.A., *Mexico*
Prentice-Hall of India Private Limited, *New Delhi*
Prentice-Hall of Japan, Inc., *Tokyo*
Pearson Education Asia Pte. Ltd., *Singapore*
Editora Prentice-Hall do Brasil, Ltda., *Rio de Janeiro*

CONTENTS

PREFACE   CHANGE!   v

INTRODUCTION ONE STEP AHEAD   vii

CHAPTER 1   FINDING YOUR WAY   1

        1.1   SEARCHING THE WORLD   Searching   1
        1.2   BEING A LITTLE PICKY   Evaluating   4
        1.3   WHAT YOU'VE FOUND   Citing   6

CHAPTER 2   NEWS OF THE DAY   9

        2.1   A SMORGASBORD   General News   9
        2.2   BY THE MENU   Science News   10
        2.3   LOST IN THE LIBRARY   Library Resources 11

CHAPTER 3   STAYING IN TOUCH   13

        3.1   A MAILBOX IN CYBERSPACE   E-Mail Accounts 13
        3.2   A PLACE TO CALL HOME   Home Pages   16
        3.3   A CALENDAR OF EVENTS   Calendars   19
        3.4   SOMETHING CALLED PRIVACY   Privacy   19

CHAPTER 4   STAYING IN TUNE   21

        4.1   A WEB COMPANION   Website Basics 21
        4.2   ANYWHERE & ANYTIME   On-line Education 24

GLOSSARY   IT'S ALL GREEK TO ME   27

DEDICATION

This new edition is dedicated to my sons, Nicholas and Nathan. They are both under 3 years of age but each has taught me patience, given me tremendous joy, and opened my mind to a new vision of the world. I look forward to every new day of discovery with them. I can only imagine what challenges and wonders the world will offer them when it's their time to walk.

This new edition would not have been possible without the continued love, encouragement, and support of my wife, Elizabeth. She is my inspiration for attempting to highlight the Internet as a tool for teaching and learning. I continue to be inspired by her inexhaustible energy and amazed by her skill for teaching and love for learning. This newest edition is in large part due to her experience as a teacher and skill as an editor. Finally, I would like to thank everyone at Prentice Hall for allowing me to explore and at times redefine the role of technology in the classroom.

Andrew T. Stull

I thank Andrew Stull and the staff at Prentice Hall for giving me the opportunity to share my classroom technology experiences with you through this new edition. During its preparation, I have enjoyed the patience, support and encouragement from my wife, Dessa. I am also thankful for numerous conversations with young people, including my own three grown children, engaged in one of the most significant technological advances in history. Global communication shaped by the Internet will change our lives.

Harry Nickla

PREFACE
CHANGE!

The first edition of this guide was completed at 7:22 p.m. on July 23, 1995, and the second edition at 2:04 p.m. on January 31, 1997. At the moment this third edition reaches completion, we'll probably record the time as well. You might think it odd that we should perform this ritual—but think again. At an earlier time in our history, information took months or years to cross continents and oceans. Any news less than a month old was "late breaking." The accuracy of news today is obviously measured differently; it now takes seconds for information to circle the globe. In fact, the popularity of the Internet has created a situation where individuals hear about important news events before the newspapers even have a chance to write about them. The world has changed, and it continues to change. Our world is louder, faster, and more complex than that experienced by earlier generations, but it also offers more promise. In terms of information transfer, we might be described as a techno-generation; our parents generation might be described as a paper-generation. Dealing with change is a basic requirement for surviving in our modern world. But fear not, our on-line future might be chaotic but it should also be exciting. Prepare yourself to revel in this change.

This manual has four chapters. The Introduction, One Step Ahead, briefly describes basic techniques and tools with which you should already be familiar. We will not go into detail on many of the topics that were covered in the first two editions of the Student's Internet Guide. There are many resources available to help beginners and novices; we will gladly point you toward a plethora of information to help you get started. This resource will help you begin to use the Internet as a tool—a tool for communicating, a tool for optimizing your workload, and a tool for navigating the jungle of information available to you. You may still choose to use the Internet as a toy also, and we'll even show you a few places to start, but this guide is meant primarily to help you face the information challenges ahead of you.

In Chapter 1, Finding Your Way, you will review techniques for gleaning information from the Web. Within this chapter you will learn about searching for, evaluating the merit of, and properly citing information. Learning to use the Internet judiciously will be a distinct asset, as it can be a channel for misinformation.

In Chapter 2, News of The Day, you will explore ways to use the Internet to stay in touch with the news of the world and with scientific happenings. The Internet, as a new media for communication, includes news sources from television, radio, newspapers, and magazines.

In Chapter 3, Staying in Touch, you will learn how to use simple resources, such as e-mail, homepages, and calendars, for managing your time and organizing your busy schedule. The Internet is primarily a communications tool. Whether it is used for direct communication, such as with electronic mail (e-mail), or whether it is used for passive communication, such as with Web pages, the Internet is becoming the most prominent way for you to gather and disseminate information to the world.

In Chapter 4, Staying in Tune, you will learn about the tools provided for you that are companions for your textbook and what on-line education has to do with you. As the digital revolution progresses, the definition of a book will change. In addition to the chapter resources found within your textbook, there are many chapter resources found within its digital Web companion and often in its CD-ROM.

Throughout this guide, you will find a collection of basic and advanced resources to help you make the most of your study of science and your study of life on the Internet.

Reading this manual won't teach you all there is to know about the Internet, but it will help you to teach yourself. If you are successful, your skills in "harvesting" information from the Internet will allow you to deal with perpetual change. When you have finished this manual, you should be comfortable and resourceful in navigating the complexity of the Internet—from its back eddies to its thriving thoroughfares.

INTRODUCTION
ONE STEP AHEAD

This thing that we now call the Internet has been evolving ever since it was first developed almost thirty years ago. Its prominence in our society has been increasing exponentially in recent years. It is unlikely that you are reading this manual without some basic understanding of the Internet and its features. Furthermore, we're pretty confident that most of you have considerable knowledge of the Internet. While earlier editions of this guide were written with no expectation of your Internet experience, this is no longer a realistic position. Numerous "Internet guides" are available to help the beginner connect to and browse the Internet. If you need a refresher then skip to the next page and you'll find a list of URL's to help you get started or to refresh your understanding. This third edition of the Internet guide will take the next logical step to help you make the most out of the Internet as a tool. Let's justify this with a bit of history. At one point in time, telephones came with an instruction manual. This was when telephones were still new, and a telephone number was just a confusing string of numbers. Today, if you see a phone number you know what it is and how to use it. Can you imagine how funny it would be to have someone think that they needed to explain how to interpret a telephone number? The Internet might not be as integrated into our culture as the phone system, but it will be and not too far in the future. For a majority of you, a string of characters such as *http://www.something.com* already has meaning. Don't worry, if you're not yet this familiar with the Internet, we're not going to leave you in the dust, but you will need to do some homework. The Web sites listed below will point you to many helpful resources about the on-line world. You'll be Web surfing in no time.

The Internet was born as the solution to a problem. It was designed to provide a global communication channel for the exchange of scientific information and research. Gradually, however, the Internet has also become a digital post office, a digital bulletin board, a digital telephone, and a digital tutor. Depending on whom you listen to, it may eventually be a digital television, a digital textbook, or even a digital classroom. The bottom line is that the Internet is growing in many directions as people realize its potential and employ its power to solve their problems. But don't get too far ahead; its real merit to you is how it will solve <u>your</u> problems and make <u>your</u> day just a bit more manageable. Hopefully, that is what you'll discover here.

For those of you with a driving quest for knowledge, here are a couple of Web addresses discussing the history, growth, and culture of the Internet. In addition, there are several basic tutorials to help you refresh your knowledge and fill in the gaps you might have about browsing, bookmarking, and connecting to the Internet.

History

History can be a valuable tool if you wish to understand the nature of things. Often, history can be used to predict the future. If you ever wanted to know why the Internet came into existence, how it has changed since its birth, or where it might go, then the following resources are the last stop you'll need to make.

BBN Internet Timeline
http://www.bbn.com/about/timeline

The BBN Timeline places important events about the Internet in context with other historical events while tossing in plenty of social commentary to give you some perspective.

JF Koh's Internet Timeline
http://wwwtds.murdoch.edu.au/~cntinuum/VID/jfk/timeline.htm

This site presents a very detailed historical perspective about the technology innovations that led up to the birth of the Internet. It begins in 1642 with Pascal's mechanical calculator.

Hobbes' Internet Timeline
http://info.isoc.org/guest/zakon/Internet/History/HIT.html

This site offers a great deal about the Internet, the people who use it, and on-line culture. It also offers some of the best information about the Internet's growth.

Getting Started

The following URL's are a few of the many beginner's guides available on the Internet. You'll find everything you need to know about modems, browsers, e-mail, bulletin boards, chat rooms, and getting connected to the Internet on at least one of these sites.

An Introduction to the Internet from Interactive Connections
http://icactive.com/guide/index.htm

This is a very comprehensive guide to the Internet. It is provided by Interactive Connections, an Internet Presence Provider. If you need a refresher course on Internet basics or are starting from scratch then this site will help.

Learn the Net

http://www.learnthenet.com/english/index.html

Learn the Net specializes in on-line training products and services for the corporate world. Their guide is well written and up to date. It is an excellent source of information for the beginner.

Net Guide from PC User Magazine

http://www.pcuser.com.au/netguide/

This guide is sponsored by the Australian version of PC User Magazine. In addition to a selection of articles about tools and techniques for experienced Web surfers, this site contains many useful resources for the beginner.

CHAPTER 1
FINDING YOUR WAY

Many of you reading this guide have a lot of experience with computers, while others have little or none. Before proceeding, you should have and be familiar with a few basic resources:

1) Computer
2) Web browser
3) Internet connection

Don't worry if you can't afford your own resources. There are many free or inexpensive options available to you, and we'll do our best to show them to you. The use of computer labs is now a common and even required component of a science curriculum. Also, we're pretty confident that these computers have one of the popular browsers by Netscape or Microsoft and an Internet connection. If you haven't found your campus computer lab yet, then our guess is that you'll find it associated with your campus library. From a beginner's point of view, the only real concern you'll have is learning the basics.

SECTION 1.1
SEARCHING THE WORLD

Although, many wire-heads consider the Internet to be the largest library on the planet, it doesn't necessarily have the easiest card catalog in the world. In this section, we'll explore techniques for searching the Internet, discuss practices for evaluating the validity of the content you find, discuss on-line education, explain CD-based Companion Website learning, and review guidelines for citing information within your class assignments. With practice, these skills will help you improve your usage of the Internet.

There is one skill, or rather behavior, that you must adopt in order to maximize your time-to-gain ratio. That is, be aware of "search drift." The Internet is an information jungle and if you wander into it without having a sound idea of why you are there or if you just wander around without being aware of where you are, then you will get "lost" and waste a great deal of time. Yes, there are times when you will want to play, wander, and have a good time, but consider whether the best time to do that is the night before a test.

SEARCHING

Yahoo! is a good place to begin. It is only one of many resources available on the Internet. It's easy to remember. If you have a chance, log onto the site and follow along as we describe how to use it.

Yahoo!:   http://www.yahoo.com

Yahoo! began as a simple listing of information by category—kind of like a card catalog. As it's grown, it has added the ability to search for specific information—and many, many other features that we encourage you to explore. At the top level of the directory, there are several very general categories, but as you move deeper into the directory, notice that the categories become more specific. To find information, you simply choose the most appropriate category at the top level and continue through each successive level until you find what you're looking for (or until you realize you're in the wrong place). Don't be afraid to experiment—it's easy to get lost but also easy to find your way home.

Suppose you were studying the impact of humans on coral reefs. Within Yahoo!, notice that one of the top-level categories is Science. Science seems as if it would be the most appropriate place to find this information, so give it a try. After accessing the Science category, you'll notice that it gives a list of many different types of science. So, what is your next choice? One choice might be Biology but Ecology or even Geology could be possibilities; in fact, Marine Biology might be the best choice. Because Yahoo! cross-references among the categories, you'll find that several related categories will lead you to your desired page.

Much of your success in finding information with this type of tool really centers around your preparation for the search. Often, it is possible to find information on a topic in a category that may at first seem unrelated to your topic of interest. Again, let's take the example of coral reefs and human impact. Although you may consider this a science-oriented topic, there are other avenues to consider. Aren't coral reefs a common destination for vacationers? Categories related to vacations and travel could be searched. Where are the coral reefs in the world? One of the largest is in Australia; by selecting Australia and neighboring countries, you might also turn up something related to coral reefs.

Prepare yourself for a search <u>before</u> you jump into one. In the long run, it will save you both time and frustration. Don't be afraid to try some strange approaches for your search strategy. A good technique is to pull out your thesaurus and look up other names for the word. You might be able to find a more common form of the word. Think of everything associated with your question and give each of these subjects a try. You never know what might turn up a gold mine of information.

The following list of resources contain many more helpful tools, tips, and techniques for searching the Internet. If you have specific academic needs, many of these tools are what you'll want to use.

Librarian's Index to the Internet
http://lii.org/

Nueva School Search Strategy Planner
http://www.nueva.pvt.k12.ca.us/~debbie/library/research/adviceengine.html

NoodleQuest Search Tools
http://www.noodletools.com/noodlequest/

Using Internet and Web Search Engines Effectively
　An Online Course from the American Library Association
http://www.ala.org/ICONN/advancedcourses.html

SEARCH ENGINES

A more direct approach to finding information on the Web is to use a search engine, which is a program that runs a search while you wait for the results. Many search engines can be found on the Web. Some Web search engines are commercial and may charge you a fee to run a search. Search engines are also available for other parts of the Internet: Archie, Veronica, and Jughead are examples of such search engines.

As mentioned earlier, Yahoo! has a useful search engine. Another search engine that is used frequently is called Lycos (http://www.lycos.com). It's simple to operate but, as with any search tool, it takes practice and patience to master. Take the time now to connect to Lycos, and we'll take it for a test run. When you first see the opening page, you'll notice that it is very complex. But it's an excellent resource, and the instructions on the page will tell you almost everything you need to know. To search, enter a word into the white text entry box and press the submit button. Lycos will refer back to its database of information and return to you a page of hyperlinked resources to various sites on the Internet that contain your search word.

To see how a search engine works, use coral reef as a topic for a search. Notice that you can set the number of responses that the engine will return to you. Did you notice that some of your results didn't seem to apply to your topic? This is one of the pitfalls of search engines. They are very fast, but they don't think—that is your job. A search using the term coral reef is just as likely to turn up a link to Jimmy Buffett's Coral Reefer Band as a link to coral reef research. To perform an effective search, you will need to spend time <u>before</u> the search preparing a search strategy. When you do research using an automated tool like a search engine, you can expect many links to be unrelated to your topic of interest—but all in all, search engines are still very powerful tools.

Another type of search service that you'll hear much about is called a meta-search service. This type of service will send your query out to a number of different search engines and then tabulate the results for you. They come in many different levels of sophistication and they also generate a large amount of information. If you're not intimidated by volume then give one of them a try.

> Here's a meta-search tool that is both fun to use and powerful. Give it a try.
>
> **Ask Jeeves**   http://www.askjeeves.com

One last word on search engines. These tools don't directly search the Internet. They actually search a database that is derived from the Internet. Here is how it works. Search engines use robots (automated programming tools) that search for and categorize information. This information is placed into a database. It is this database that you search when you use the search engine. Can you think of a potential problem with this system? Unfortunately, the quality of the database depends on the effectiveness of the robot that assembles the database. This is why you should not rely on just one search engine tool. Use several because what one does not find, another might. You shouldn't have trouble finding other search engines if you don't like the ones we list here. Both of the major browsers now include a basic menu button that will connect you to a large collection of different search engines.

> The following resources will help you learn more about searching the Internet.
>
> **How To Search the Web from Palomar College**
> http://daphne.palomar.edu/TGSEARCH
>
> **Search Engine Watch**
> http://searchenginewatch.com
> http://searchenginewatch.com/resources/tutorials.html
>
> **Search Guide**
> http://www.searchengineguide.org
>
> **Meta-Search Engines Guide**
> http://www.hampton.lib.nh.us/srchtool/recmetaengines.htm

SECTION 1.2
BEING A LITTLE PICKY

In your career as a student and eventually as a professional, you will spend a great deal of time using the Internet to communicate and find information. But can you trust the information you find? In traditional publications, just as with Internet publications, there are strong, reliable sources of information and then there is the other end of the spectrum. Is it possible to leave a grocery store without passing a tabloid newspaper displaying a title like, "Elvis Was A Spy for Alien Invaders?" It's obvious that this title is misleading—we all know that Elvis was actually a double agent and on our side. When information is outlandish, it is easy to spot the truth from the lie, but not everything is as obvious. Misinformation is occasionally passed on by respected publications as well. Developing skills to evaluate all sources of information intelligently, especially those from the Internet, will be a valuable asset.

Evaluating the merit and accuracy of an information source isn't new to the Internet. Criteria for evaluating the Internet have been adapted from the previous standards to reflect the unique challenges offered by the speed and global nature of the Internet.

The following is a list of criteria that can be used to evaluate information sources. It is adapted from traditional evaluation criteria and personal observations of the behavior of Internet authors, managers, and publishers. Additionally, there are many Internet sites that discuss evaluating information from the Internet. You may eventually develop additional criteria by which you evaluate Internet sources, but these should get you started:

1. **Authority**
   Who is the author and what are his/her credentials?
   Does the source cite respectable references and have a contact point?
2. **Accuracy**
   Does the piece follow basic spelling, grammar, and composition rules and appear to be reliable?
   Are the embedded hyperlinks valid and do they present accurate information?
3. **Objectivity**
   Is the source biased or does the author have an agenda?
   What is the purpose or intent of the piece—persuasive or informative?
4. **Currency**
   When was the original piece created and is it updated regularly?
5. **Coverage**
   Who is the intended audience?
   How comprehensive is the piece and how well is the topic covered?
6. **Stability**
   Is a student, an instructor, or an institution the author of the piece?
   Is the URL likely to change over time and what is the domain?
   What is the primary focus of the organization behind the sponsoring domain?
7. **Utility**
   Is the piece valuable as a source of primary information for a topic?
   Is the piece valuable as a source of reference information for a topic?
   How is this piece used (e.g., reference, service, communication)?

If you wish to review criteria that others have proposed for evaluating Internet resources, look through these sites. Each will offer you a unique point of view, but all are valuable sources of information that we encourage you to read.

Evaluating Quality of the Net from Babson College
http://www.tiac.net/users/hope/findqual.html

Evaluating Internet Resources: A Checklist from University of California, Berkeley
http://infopeople.berkeley.edu:8000/bkmk/select.html

Evaluating World Wide Web Information from The Libraries of Purdue University
http://www.lib.purdue.edu/StudentInstruction/evaluating_information.html

Ten C's for Evaluating Internet Resources from University of Wisconsin-Eau Claire
http://www.uwec.edu/Admin/Library/Guides/tencs.html

Thinking Critically about World Wide Web Resources from University of California, Los Angeles
http://www.library.ucla.edu/libraries/college/instruct/web/critical.htm

SECTION 1.3
LISTING WHAT YOU FOUND

The next logical step after you've found and evaluated your sources of information is to correctly cite them in your work. Citing scientific references requires a specific format. A natural part of all scientific reports is a citation list that supports the background, design, and conclusions of the described research. Different scientific disciplines and journals will have different formats but in general, there is a common design by which scientific references are cited. For example, the Council of Biological Editors has a standard style for cited references. As we've seen with the other topics included in this chapter, the traditional way of doing something often needs to be modified to include on-line content. It is a simple matter of properly citing on-line references so we will not go into it here. A number of print and on-line reference works are available to help you. Additionally, many journals will have their own format for listing on-line references. Remember, it is important to properly evaluate your references so that they have merit as a cited work.

The following list of URL's are sources to help you learn the proper way to cite on-line references.

Citation Style Guide from West Chester University
http://www.wcupa.edu/library.fhg/resource/citing.htm

Citing World Wide Web Information from The Libraries of Purdue University
http://www.lib.purdue.edu/StudentInstruction/citing_sources.html

Columbia Guide to Online Style from Columbia University Press

http://www.cas.usf.edu/english/walker/mla.html

Electronic Reference Formats Recommended by the American Psychological Association
http://www.apa.org/journals/webref.html

Guide for Citing Electronic Information from William Paterson University
http://www.wilpaterson.edu/wpcpages/library/citing.htm

Online! A Reference Guide to Using Internet Sources from St. Martin's Press
http://www.smpcollege.com/online-4styles~help

CHAPTER 2
NEWS OF THE DAY

The Internet is proving to be a fast means of distributing news to the world. Even before the news agencies have a chance to print the news, it is available to you on the Internet. Your problem is not one of news access but of news volume. You may still be asking why you should be concerned with science happenings. The simple answer to this question concerns power, control, and security. If you want to be able to decide how the world around you changes, what medical options are best for you, and the impact of pending legislation on you and your family, then knowledge of science is a must. The fastest track to awareness of scientific events is through the news. For the pragmatic among you, it is also a great source of information for reports and current event assignments.

In this section of the chapter, we'll explore a few of the many sources of general and science news on the Internet, and you'll see how easy it can be to surf the Internet without being buried by the information wave.

SECTION 2.1
A SMORGASBORD

Science is an everyday tool used by which our society addresses and challenges our modern age. With the prominence of the Internet, it is very simple to stay on top of the happenings and events in science so that you can influence your life's course.

The following list of organizations offer convenient access to science news. The URL listed for each news source is the general Web address for that source. Upon reaching one of these Web sites, you can choose a number of different categories of news. In a majority of the cases, "Environment," "Health," "Science," "Technology," or "Sci/Tech" are categories that will lead you to news with a science focus.

> Often, if you see or hear of a news program on television, you can go to their Internet page for details and related resources.

Television	Web Address
ABC News	http://abcnews.go.com
BBC News	http://news.bbc.co.uk
CBS News	http://cbsnews.cbs.com
CNN	http://www.cnn.com
Discovery Online	http://www.discovery.com
FOX News	http://www.foxnews.com
MSNBC	http://www.msnbc.com
PBS	http://www.pbs.org

Occasionally, some of these news services may require you to register to receive their news but most are entirely free. The common model for these Internet news sources is similar to that of commercial television. You've first got to wait through the commercials before you can watch your show of interest.

In addition to the popular news shows on television, newspapers and radio both offer you more in-depth coverage through their Web sites. In many cases, it is also possible to search their archives for past articles. The following is only a very small subset of the newspapers and radio stations available to you.

Newspapers/Radio	Web Address
BBC World Service	http://www.bbc.co.uk/
LA Times	http://www.latimes.com
NPR (National Public Radio)	http://www.npr.org/news/
NY Times	http://www.nytimes.com
Science-Friday	http://www.sciencefriday.com
Seattle Times	http://www.seattletimes.com
USA Today	http://www.usatoday.com

SECTION 2.2
BY THE MENU

Although television and newspaper Internet sites may offer interesting and informative articles, they probably don't offer the detail or emphasis that is at the heart of a scientific publication. For this reason, you may find that you need a science-related source of information without the political and social commentaries. The following list of sites are an excellent place to begin your search for this type of information.

Many are free but a few require you to subscribe to their service. (Some of these science news sites may require a subscription fee.)

Science Information	Web Address
American Scientist	http://www.sigmaxi.org/amsci/amsci.html
exoScience	http://exosci.com
InScight	http://www.academicpress.com/inscight/
NASA	http://www.nasa.gov
Nature	http://www.nature.com
Newswise	http://www.newswise.com
Quadnet	http://www.quad-net.com
Science	http://www.science.com
ScienceDaily News	http://www.sciencedaily.com/index.html
Science News	http://www.sciencenews.org
Scientific American	http://www.sciam.com
Unisci	http://unisci.com

A few of these are peer-reviewed research journals for professional scientists; therefore, they may be quite expensive. If you need an article from one of these on-line journals, your library may have a site subscription or one of your professors may have a research subscription. Don't open your wallet before you try all your options.

The last thing you should remember about on-line news services is that most of them offer a search and archiving tool. Often, their articles are aggressively hyperlinked to related materials throughout the Internet. Although you won't see something of interest at the top of the page, you might find something interesting with a quick search. Further, an article of average interest might take you to something truly valuable with one of their embedded links.

SECTION 2.3
LOST IN THE LIBRARY

In case you thought your options were now exhausted, consider the following. Even in the Internet age, your library is still the best source of information. Your campus library has access to many networked tools and databases that are not directly available on the Internet. For example, PALS is a common on-line card catalog used by many libraries. It allows users to search library holdings by author, title, term, etc. When searched, this automatic card catalog returns a "record" providing the location of the holding in that particular library if it is present. A natural offshoot of this is something called WebPALS— by now you should be getting the picture. WebPALS is like PALS but it is an Internet application that allows a connection, through the institution's library, to tens of hundreds of libraries. Your library may or may not subscribe to this particular service but it is likely to have something similar.

While your on-line card catalog can help you find <u>where</u> the information is that you want, the missing piece is the tool to help you get the actual information. One of the more likely services that you'll use, and use often, is the Interlibrary Loan program. Like the shared on-line catalog, libraries also share their information with other libraries. In this way, you can borrow materials that you library does not actually possess. Your librarian can give you more information.

In addition to your ability to search and retrieve books from distant libraries, other services are also available to help you search and retrieve information from other sources. Through your library you have access to a tremendous collection of databases. First-Search is a common tool that you're like to use to both search and retrieve information from these databases. It is a gateway system that provides access to dozens of databases. Those most commonly used in the sciences include:
- *ArticleFirst:* broadly includes humanities, social sciences and sciences
- *BasicBIOSIS:* covers 10 years of Biological Abstracts and core journals
- *ContentsFirst:* covers 12,000 journals

- *Dissertation Abstracts:* full coverage of dissertations back to 1861
- *Electronic Collections General Abstracting Resource:* heavy in science and medicine
- *General Science Abstracts:* 140 science titles
- *Government Publications:* indexing back to 1976
- *MEDLINE:* Medical Journals
- *PAIS:* indexes articles, books, conference proceedings, government documents
- *PapersFirst:* indexing of conference papers held at the British Library
- *WorldCAT:* holdings in the Library of Congress and research libraries

An on-line card catalog and FirstSearch are only two of the tools your library is likely to have. In general, these and other tools are free through the local network, however off-campus use may require a password. Check with your librarian for help with specific procedures.

CHAPTER 3
STAYING IN TOUCH

Although the Internet is sometimes thought of as a flashy, graphically rich waste of time, it began as a tool to enable researchers to communicate between research labs across the United States. If you look at its basic features, the Internet is still a valuable and effective tool for communication In essence, one goal of the Internet has been to eliminate the hindrance of geography on the free exchange of ideas. Whether it becomes a waste of time or a time-saving tool is entirely up to you. We hope the following ideas will help you make the most of the Internet as a tool for communication and collaboration.

SECTION 3.1
A MAILBOX IN CYBERSPACE

An e-mail account is the most basic of methods for planting yourself in the Internet community. Do you have one? Don't worry if you don't. We have a number of simple, inexpensive, and fast solutions you may want to consider.

There are a few options available to you. You may be able to apply for an e-mail account through your college. If your college doesn't provide student e-mail accounts, then e-mail service through an Internet Service Provider (ISP) is a second option. ISP's require you to subscribe (meaning spend money) to acquire their service. The nature of service, hourly or monthly, will depend on your anticipated use. The disadvantage is that you will need to pay a fee for the service. This can also be considered an advantage because you can expect help from time to time, which you are not as likely to receive from other options.

Should you wish to pursue this option—and if you have the cash—you can find a national list of ISPs at the following address: (http://www.boardwatch.com). Costs average about $20 per month depending on the services that you use. We suggest that you do not sign a long-term contract with an ISP until you are certain that you are happy with the service that particular provider offers. Most providers offer a free trial period before any formal commitment is necessary. Test the system at various times during the day to be certain that sufficient access is provided.

A third option, which is increasing in popularity, is to choose a free e-mail service provided by one of the many on-line companies. Yes, a free e-mail account with many of the bells and whistles found in a regular e-mail account can be yours for the asking. If you choose a free e-mail service, then read the fine print and understand what it means to you. In most cases, the service is provided to you free because the provider is making its money by selling advertising space to other companies. This is the same way that search engine companies and television stations make their money. In order to read your mail, you have to wade through a few commercials prominently posted on your e-mail reader. An additional condition of these free e-mail accounts is that they will gather information about you in order to customize and target the display of commercials for you. In most cases, this information is used only to target you with commercials but always read the fine print.

The following are only a few of the more prominent services offering free e-mail and free Internet access in general. Read the fine print in their service agreements, and choose the one that offers you the most. Also, don't be afraid to change services if you're not getting what you expect.

E-Mail Service	Web Address
AltaVista	http://www.zdnet.com/downloads/altavista/
BlueLight	http://www.bluelight.com
Hotmail	http://www.hotmail.com
Juno	http://www.juno.com
Netscape	http://webmail.netscape.com
NetZero	http://www.netzero.com
WorldSpy	http://www.worldspy.com/freeisp/isp.html
Yahoo!	http://mail.yahoo.com

So, now that you're on your way to your own e-mail account, what are you going to do with it?

Simple Suggestions

If you wish to skip all of the instructions, here are a few suggestions to keep you out of trouble.

Write down the user ID and password for your account. It's difficult to read your e-mail if you can't get into your account.

Change your password periodically. Someone stealing your login information could do a number of unscrupulous things with your account and reputation.

Don't use the same password for all of your accounts. Yes, it is much easier to remember if you do but it is also much easier for someone else too.

Watch out for e-mail viruses. They are common and can unintentionally be passed through attached documents.

E-Mail and Your Instructor

E-mail is becoming a very common and popular way for students and instructors to communicate outside of class. As you progress through college, it is likely that you will have numerous e-mail exchanges with your instructors. The following should help you greatly.

- When communicating with your instructors, use correct spelling, grammar, punctuation, and clarity—just as you would with a carefully crafted letter.
- Most instructors will refrain from sending confidential information through e-mail since one can't guarantee the security of the message. Therefore, it is best not to request confidential information, exam scores, or course grades electronically.
- If you are asked to submit assignments electronically, be very careful as to the timing and the format you select.
- Smaller bits of text, such as summaries or project descriptions, can be sent in the body of the message; however, larger documents, including graphs and tables should be sent as attachments. Your instructor will give you specific instructions about submitting such documents.
- Most instructors will have a mechanism for acknowledging receipt of important documents. If you have not received an acknowledging document, be certain to check by phone or in person with the instructor. It is the student's responsibility to be certain that all assignments are received in an acceptable form.

E-Mail Etiquette

Etiquette is especially important with e-mail communication. When engaged in a conversation, it is likely that you are also communicating information with the inflections in your voice, the expression on your face, and the posture of your body. If you take any or all of these away, there is a greater chance for miscommunication. Here are a few suggestions to help you out in the e-mail world.

- Say what you mean, say it concisely, and say it very carefully—once you've sent it, it is "there" and cannot be retrieved. We have all had to follow-up a vague or hurtful e-mail with explanations or apologies.
- Get to the point—your instructor is probably very busy and will be unwilling to read a tome. If you want to chat then we suggest a pizza.
- Use the subject line—it's a quick way to tell the other person what you want.
- Don't shoot from the hip. Sometimes normally timid people become raging bulls when on-line.
- Understand the distinction between Reply and Reply All on the menu bar—or you may have just sent your most passionate love-letter to a mailing list.
- Use a "smiley" when you think there is a possibility for misinterpretation—with e-mail, there is no opportunity to convey varied meanings by tone of voice or body language.

There is something to be said for visual communication. Let's say a professor sends an email to schedule a meeting. Then, let's say a colleague replies to that e-mail with a message that simply says "woopee." Is it a sarcastic response or a light-hearted

15

verification? A simple smiley would have helped to communicate the writer's intention.

> To help you get started and to make you look more like a seasoned expert, here are a few common smileys to start your collection.
>
> | | |
> |---|---|
> | :-) | Smile, nothing serious intended |
> | ;-) | Winking smile, perhaps flirtatious or sarcastic |
> | :-(| Frown, something is bothering the author |
> | :-I | Indifference |
> | :-> | Possible biting or very sarcastic remark |

This is by no means all there is to know about etiquette on the net (nettiquette) and the ins-and-outs of e-mail but it's a beginning. Each institution will supply more definite guidelines. Read them and follow them. The following list of URL's should help you find, understand, and use your e-mail to peak efficiency—or at least to maximum entertainment.

> A Beginner's Guide to Effective Email by Kaitlin Duck Sherwood
> http://www.webfoot.com/advice/email.top.html
>
> **Optimized E-Mail from 1 2 3 Promote!**
> http://www.123promote.com/workbook/plan1.htm
>
> **Email Etiquette from Air Canada**
> http://www.acra.ca/mlist/emailetiquette.htm
>
> **ICONnect Online Courses on E-Mail from the American Library Association**
> http://www.ala.org/ICONN/ibasics2.html

SECTION 3.2
A PLACE TO CALL HOME

After setting up an e-mail account, a home page is the next logical step toward establishing yourself with an Internet presence. Considering the proliferation of personal home pages and the typical merit of their content, you might not realize the advantages that a personal home page may offer to you. While an e-mail account offers you an identity on the Internet, a home page offers you a central resource that is mostly under your control. As a student, you are somewhat nomadic and therefore required to work in many different locations throughout the day. A home page can be an important central resource for your nomadic life. Your home page could list on-line reference sites such as search engines, dictionaries, directories, and glossaries; a hyperlinked list of e-mail addresses for your instructors, classmates, and friends; a place where you can post shared information for your study groups; or a place to post class assignments for your instructors. In short, a home page may be passive in nature but it can be a valuable tool for communication and it can save you a lot of time.

You have three basic options for posting and maintaining a home page on the Internet. Your college may offer you space to post and maintain a home page, you can subscribe to an ISP, or you can use a free service. The business model used by free e-mail services is similar to that of companies providing free home page services. In most cases, these services have a basic format that you can occasionally add to or modify. Read the fine print to make sure you understand the agreement.

The following services enable you to set up a home page on the Internet. Each of them offers a slightly different service, so spend a bit of your time to really evaluate their offerings.

Home Page Service	Web Address
Geocities	http://www.geocities.com/join/
Microsoft	http://home.microsoft.com/
Netscape	http://my.netscape.com/
Yahoo!	http://my.yahoo.com/
NetColony	http://www.netcolony.com

Part of the fun of having a home page is creating it to reflect your interests and personality. As you begin moving through the Web you'll notice a great variety of home pages. Some of them are not so good but a number are both expressive and functional. As you begin to design your own home page, remember what you want it to do and say about you.

The following list of on-line resources should help you begin building your first home page. With a quick search of the Internet, you will find a large number of other resources along this line. Be creative and enjoy the experience.

A Beginner's Guide to HTML
http://www.ncsa.uiuc.edu/General/Internet/WWW/HTMLPrimerAll.html

The Bare Bones Guide to HTML by Kevin Werbach
http://werbach.com/barebones/

If this isn't enough then the following sites can give you even more information on Web page design:

Internet.com	http://www.webreference.com
Art and the Zen of Web Sites	http://www.tlc-systems.com/webtips.html
Creating Killer Sites	http://www.killersites.com
WebDeveloper.com	http://www.webdeveloper.com
Web Building	http://builder.com
Web Monkey	http://www.hotwired.lycos.com/webmonkey

The Need for Plugins

Plugins are software programs that extend the capabilities of a particular browser in some specific manner, giving you the opportunity to play audio samples or view movies from within the browser. Such Plugins are usually "cross platform" in that they can be used on Macintosh or Windows systems. Below are some examples of important and popular Plugins that you'll probably need to view the more interactive Web sites.

- *Flash Player* by Macromedia—This plugin will allow you to view animation and interactive content through your browser. This interactive content includes cartoons and games from leading-edge companies like Comedy Central, Sony, and Disney. You'll also need this plugin to view many of the science animations being developed for your books. (http:www.flash.com)
- *Shockwave* by Macromedia—This is the industry standard for delivering interactive multimedia, graphics, and streaming audio on the Web. Major companies like CNN, Capitol Records, and Paramount use Shockwave as their delivery system. (http://www.macromedia.com/shockwave/download/)
- *RealPlayer* by RealNetworks—This plugin allows you to play audio, video, animation, and multimedia presentations on the Web. RealPlayer Plus gives sharp pictures and audio for RealAudio and RealVideo. Many popular radio and televisions shows are available on the Web if you have this plugin. (http://www.realplayer.com)
- *QuickTime* by Apple Computer—This plugin allows you to play audio/video productions and is commonly included on CDs. It is extremely common and typically preinstalled in the recent version of both major browsers. Upgrades are very frequent so you can always download the newest version at their Web site. (http://www.apple.com/quicktime/download/)

If you wish to download these or other plugins then you can go directly to the company that makes them or to the download gallery of the browser that you use. Both major browsers provide a listing of plugins by category for your access. Simply download the one you want and then follow the installation instructions.

SECTION 3.3
A CALENDAR OF EVENTS

The final step in our project to help you stay in touch with your classes, friends, and family is to make you aware of the help that a calendar program can lend. By now, it should be obvious to you that your life is not going to get less complicated. Having a tool to help you schedule your time and remember important events will be a distinct asset. Developing a routine to organize your life is the first and best step to take. The second step is to find a tool to help you remember your reading and homework assignments, library time, class schedules, exams, study group meetings, and office hours for your instructors—in addition to all of your personal commitments.

As always, your options are numerous. Memory is probably the most common option initially employed by the novice, but its disadvantages are obvious. There are also paper calendars specifically designed to help you organize your time. Let us lead you to yet another of those free on-line resources for a final option. It is not necessarily your best option, but it has advantages. In addition to offering you free e-mail, Yahoo!, Netscape, and many other companies offer a free on-line calendar service. The service is free—provided you register. By now you should be familiar with the model. The service is free to you, but you'll need to provide them with some basic personal information and you will need to endure the targeted commercials embedded in your calendar viewer. With their service, you will be able to populate a calendar with events that are important to you. Your calendar of events is viewable by day, week, month, or year. It will contain both a "To Do" list and a regular daily schedule. One of the potentially valuable resources is that of scheduled e-mail notes to remind you of important events. Never again do you need to suffer those nightmares of forgetting an exam. However, you do need to make the commitment to maintain the accuracy of your calendar. Additionally, if you know basic HTML, you can schedule events to include hyperlinks. These could be to references sites, assignments posted by your instructors, or to resources posted on the Companion Web site for your textbook. Essentially, your calendar can be completely linked to the Internet.

Parting advice: Remember that you have the power of the purse; therefore, always look for the least expensive option, read and understand an agreement before you sign it, and enjoy your journey.

SECTION 3.4
SOMETHING CALLED PRIVACY

The Internet has been moving over the last few years to increase the level of personalization that users experience when they browse. This personalization can be both good and bad—with increased personalization there is also less privacy. Information about you and your viewing and possibly purchase habits is a commodity that companies want. If your likes and dislikes are known, then it is much easier to specifically market a product to you. For example, if I notice that you always come into my record store and browse through the Blues section, it is unlikely that I'll sell you something from the Sex Pistols. But I have a better chance of pulling you in if I run a special on BB King. Some companies are in the business of providing information about you and they collect this information on the Web. Read the fine print before you sign up, register, or provide your personal information to anyone or anything on the Web.

There are plenty of powerful people worried about privacy on the Internet, so we are not alone. The following Web addresses are for some of the many organizations that have dedicated themselves to securing and lobbying for Internet privacy. It might be a helpful exercise to make a visit to their site and learn more about the situation.

Privacy Organization	Web Address
Electronic Privacy Information Center	http://epic.org
Electronic Frontier Foundation	http://www.eff.org
Center for Democracy & Technology's Operation	http://opt-out.cdt.org
Junkbusters	http://www.junkbusters.com

DoubleClick is one of the leading companies that gathers information about people and their browsing habits. If you want to learn more about what they do and how to remove yourself from their observation, visit their site.

DoubleClick   http:www.doubleclick.net/privacy_policy/privacy.htm

---

We're not trying to create a sense of mistrust in you about using the Internet. We don't want you to confuse the Internet with the X-Files. The great majority of times that you are asked to supply information on the Internet, it is safe to do so and is meant to help the requester more fully service your needs but it pays be informed about the issues involved.

CHAPTER 4
STAYING IN TUNE

Change is a central theme of science and of this guide. As you progress through college you will continue to recognize that change is also a significant aspect to all material things. You may notice that textbooks are not immune to this phenomenon either. Not that long ago, books were tremendously expensive because they were individually hand written and hand illustrated. As such, they were great works of art owned by the elite but very poor resources for reaching the masses. As is typical of our species, problems demand solutions, and the printing press was invented—probably the most significant bit of technology to have appeared in the last millennium, enabling the mass production of affordable books.

This last chapter will describe a few of the textbook-specific Internet resources that are available to you and will suggest some ideas about on-line education. Hopefully, you will see a glimmer of the future of information, education, and books through your experience with these resources. Your textbook now includes an added digital tool chest, an alter-ego if you will, called a Companion Web (CW) site and in some cases a textbook-specific CD-ROM. They contain many tools to help you visualize, communicate, and discover concepts introduced in your paper textbook. In the future, we will see the definition of a book expand to include digital content as well as the traditional content more conveniently delivered with ink and paper.

SECTION 4.1
A WEB COMPANION

The obvious question is "How do I find the CW site that goes with my book?" Fortunately, there is an easy answer to this question. All Prentice Hall textbooks have a convention for addressing their CW sites. The last name of the first author of the textbook is used to distinguish one site from another. For example, if you are in an Ecology class you are probably using "Ecology: Theories and Applications" by Peter Stiling. If you add Stiling's name to the standard Prentice Hall Web address (http://www.prenhall.com/), then you will find the CW for his book.

> http://www.prenhall.com/stiling/

In this fashion, you should be able to find a CW for any book if you know the author. Because you might not know the author for a book on every subject you'll want, there is an indirect way to reach all of the CW sites. Simply load the Prentice Hall home page (http://www.prenhall.com/) and select the Companion Website Gallery option from the page. All of the Prentice Hall CW sites are organized by discipline and are available from the CW Gallery.

Here is a partial list of science CW addresses to give you a perspective of the breadth of resources available to you.

Science Topic	Web Address
General Biology	http://www.prenhall.com/audesirk/
Astronomy	http://www.prenhall.com/chaisson/
Physics	http://www.prenhall.com/giancoli/
Genetics	http://www.prenhall.com/klug/
Anatomy and Physiology	http://www.prenhall.com/martini/
Environmental Science	http://www.prenhall.com/nebel/
Earth Science	http://www.prenhall.com/tarbuck/

Although we've only listed science CW addresses, there are CW sites that support every major discipline that you'll encounter in college. In addition, if your professor is not using a book with a CW site, then you can always browse through the CW Gallery until you find a resource to help you with any topic.

Now that you know where you can find your CW, here's what you'll see when you look inside. Just as a tool box is a container for tools, so too your CW is a container for unique Web tools. All CW's share a few basic tools accessible from the first page of their site. The CW site for Peter Stiling's Ecology textbook is a good example. In addition to an image of the textbook, you will find the following three features.

Syllabus Manager

This tool both allows your professor to create and manage an on-line syllabus for your course and enables you to view the syllabus as part of your CW. This tool is valuable to you only if your professor first develops an on-line syllabus with it. If they haven't but you think that it would be helpful, then you might consider offering to help them. You can learn more about this tool by taking the Syllabus Creation Walk-Thru found on the Syllabus Manager page.

An on-line syllabus will enable you to reach assigned activities within the CW site, sites for assignments outside the CW, and your professors individual Web pages. If your professor already has an on-line syllabus in the Syllabus Manager tool then you only need to find it to use it. To do this, simply use the search selector on the Syllabus Manager page, type in your professor's name or your college, and select the "Search Now" button. A list of on-line syllabi will be displayed from which you can select your course. Once you select your course, a calendar will appear in the lower left corner of the browser page. This calendar will contain dates and assignments as posted by your professor. It's actually fairly easy to use, but if you need more help you can select the Help tool also found on the front page of every CW site. We'll explain more about this later.

There is another scenario. If you are using a computer in a campus lab, you may occasionally find that other students will leave a syllabus loaded when you sit down to use the computer. The last option at the bottom of the syllabus calendar will unload the last syllabus and return you to the menu where you can select your own syllabus. Additionally, even if you sit down to a computer without a loaded syllabus, within the syllabus request window, you may see a list of syllabi for other courses and professors. If you've already used this tool, you might even find that your course is included in this list. You have the option of selecting your course from the list or searching for it again.

Your Profile

If you haven't done so, please read the section in Chapter 3 that deals with your privacy. This tool is intended to help make your use of the CW resources more rewarding. The information that you enter into this tool is used to help you customize your experience with the CW site. Once this information is entered, when you come back to this site, some of the basic features will be preset and ready for your use. As we mentioned earlier, be aware of the situation where someone else might use information you preset or where you might use information they preset in a shared computer. If you do share a computer, then you should probably skip over this tool.

Help

It's all in the name. If you are having a difficult time using one of the chapter tools, configuring your browser, or using the Syllabus Manager, then you can find help here. Probably the most helpful of the resources you'll find in Help is called the Browser Tune-Up. It is a resource that will diagnose your browser and its plugins to determine if you have the latest versions of the software. If you do not have the latest versions, you can download them through the tool and test them to make sure they are working properly. It will probably be helpful to do this on a periodic basis in the event a new version of the software is released.

Select a Chapter

In addition to general tools, you are likely to find a few discipline-specific tools. The chapter selector is the doorway to the meat of the CW site. All of the tools described up to this point are general and present in all CW sites, but the resources you'll find within each chapter of the CW will match the topic and intent of the chapters in your textbook. This aspect of the CW is organized in parallel to the textbook. Notice that you can navigate anywhere in the CW with the navigation bar on the left side of the CW window. This vertical bar also lists all of the tools that are available to you within the chapter. As you may have discovered in reviewing the Help tool, all aspects of the CW are thoroughly described there. We encourage you to review the different tool descriptions and to play around with each in the chapter area of your CW. Each of the tools that you'll encounter within a chapter is designed to help you understand the topics presented in the chapter. We would like to make a special note of the Feedback feature. The Internet is a dynamic place. It is not just different every day but also every second of every day. If you find something that you think will be helpful to others taking this course, that you feel is a mistake, or that could improve any of the CW sites, please send in feedback and the site will improve for everyone.

For a limited time following the purchase of a text you usually have access to technical support from the developing company. This resource can provide valuable information for technical problems that might surface. The e-mail, web_tech_support@prenhall.com will direct you to a support staff that will try to answer questions within a 24-hour period. In the current software environment, it is virtually impossible to develop highly interactive materials that will run flawlessly on all browsers but most material with work with Microsoft and Netscape browsers. Ask for help if you need it.

SECTION 4.2
ANYWHERE & ANYTIME

People all over the world are gathering to finish up degree requirements, take an elective course, retake a failed course, self-educate, or self-indulge. In most cases, people involved in on-line learning, or distance education as it is often called, are typically non-traditional students taking advantage of the convenience that distance education provides. In such a course, assignments are posted by the instructor, often on-line, then debated, researched, completed, and submitted in electronic format by the student. The instructor often communicates through e-mail, bulletin boards, or synchronous chats in what might be considered a "cybercafe" atmosphere.

WebCT in a Nutshell

It is likely that you may have one or more classes that use a powerful and common technology called WebCT (Classroom Technology). WebCT is a tool developed at the University of British Columbia to provide Web-based educational environments. It is used in over 800 colleges and universities in more than 40 countries to serve 3.6 million students in approximately 97,000 courses. It is so popular because it offers a variety of interactive functions and requires a minimum of technical expertise. WebCT courses include the following tools: a conferencing system, group presentation areas, a synchronous chat system, and electronic mail. Prentice Hall provides many WebCT resources to support its textbooks so chances are that you'll have an opportunity to explore these various tools.

A WebCT course works much like the following. A web server on campus runs the WebCT software and is used both for course creation and delivery. Students interact with course content provided by the course instructor. Students are usually assigned a password and ID for entry into WebCT. In some cases, the instructor may invite anonymous participation. Customized and optimized by the instructor, WebCT courses are just one way that you'll see the digital world revolutionizing your classroom.

On-Line Education

If you're considering an on-line course, you should read the following as a preparation. Typically, on-line courses provide a course packet, which includes the following:

- Course Study Guide (reading, writing, and/or viewing assignments, etc.)
- Textbook, Kits, Audio/Video cassettes
- Preaddressed envelopes for submitting materials, tests answers, etc.
- Detailed description of the course requirements
- Timeline for the completion of assignments
- Methods by which examinations are given

There are several critical issues that students should consider before enrolling in an on-line course:
- How will this course help you? If the course is for self-enrichment and does not involve college credit, then there is little opportunity for regret. However, if you need to receive college credit, there are several additional issues that must be considered.

If this course is not at your school, then
- Has your institution accepted the particular course? It may not be an acceptable replacement for the locally available course. Typically, an institution will not allow off-site study for locally available courses.

- Is the institution offering the distance course fully accredited? If not, it is unlikely that an accredited school will accept courses offered by that institution.

- How will the distance course fit into your progress toward your degree? Typically, distance courses are not allowed in one's major field of study.

- What is the reputation of the institution offering the course? If a major university offers the course, then it is likely that administrators at your home institution will view the course favorably.

- Do you have consistent access and knowledge to use the required technology for the course? A computer crash halfway through a semester can be disastrous, not to mention expensive.

- What are your motivations for taking an on-line course over a traditional one? If it's because you think it's going to be easier then you should reevaluate your motivations. In most cases, you'll be expected to do as much or more work.

• Do you have the motivation and discipline to complete the course if there is no formal schedule? An on-line course is not for everyone. You might be the kind of person that needs a regiment.

If you're still interested in an on-line course then one way of determining the availability is to seek the advice of administrators at your home institution. It is not uncommon for valid and viable paths to have been blazed by other students in a similar situation. Often, schools have a "college of continuing education or life-long learning" that has considerable experience with distance learning. Second, you may merely enter the term "distance learning" or "distance education" into any one of many search engines (Netscape, Yahoo!, Infoseek, etc.). Such a search will generally deliver both academic and commercial web addresses, which can be researched. The on-line education directory at http://www.caso.com may be helpful. Finally, it is quite possible that your institution is developing and offering on-line options to many of the traditional courses. It might be as simple as reviewing the course catalog.

GLOSSARY
IT'S ALL GREEK TO ME

Access Provider

A company that provides access to the Internet or a private network for a fee. (See Internet Service Provider.)

Agent

A type of software program that can be directed to automatically search the Internet or perform a specific function on behalf of a user. Spiders and worms, which roam the Internet, are the most common types of agents.

Anchor

An HTML tag used by a Web page author to designate a connection between a word in the text and a link to another page. (See HTML, Tag, and Link.)

AVI

This stands for Audio/Video Interleaved. It is a Microsoft Corporation format for encoding video and audio for digital transmission.

Backbone

The main network cable or link in a large internet.

Bandwidth

The capacity of a network line to carry user requests. Network lines such as a T1 are larger (have a higher bandwidth) and can carry more information than a lower bandwidth line such as an ISDN or a modem connection. (See ISDN and Modem.)

Bookmark

A list of URLs saved within a browser. The user can edit and modify the bookmark list to add and delete URLs as the user's interests change. Bookmark is a term used by Netscape to refer to the user's list of URLs; Hotlist is used by Mosaic for the same purpose. (See Hotlist and URL.)

Browser

A software program that is used to view and browse information on the Internet. Browsers are also referred to as clients. (See Client.)

Bulletin Board Service (BBS)

An electronic bulletin board, it is sometimes referred to as a BBS. Information on a BBS is posted to a computer where people can access, read, and comment on it. A BBS may or may not be connected to the Internet. Some are accessible by modem dial-in only.

Cache

A section of memory set aside to store information that is commonly used by the computer or by an active piece of software. Most browsers will create a cache for commonly accessed images. An example might be the images that are common to the user's home page. Retrieving images from the cache is much quicker than downloading the images from the original source each time they are required.

Chat room

A site that allows real-time person-to-person interactions.

Client

A software program used to view information from remote computers. Clients function in a Client-Server information exchange model. This term may also be loosely applied to the computer that is used to request information from the server. (See Server.)

Computer Virus

A program designed to infect a computer and possibly cause problems within the infected system. Viruses are typically passed from user to user through the exchange of an infected file. Numerous virus checkers or scanners are available to help you identify and inoculate your system against viruses.

Compressed file

A file or document that has been compacted to save memory space so that it can be easily and quickly transferred through the Internet.

Cookie

A small piece of information given temporarily to your Web browser by a Web server. The cookie is used to record information about you or your browsing behavior for later use by the server. For example, when you visit an on-line bookstore, a cookie will probably be passed to your browser to record book selections you make for purchase.

Cyberspace

This refers to the "world" of computers. It was coined by William Gibson in the novel *Neuromancer*.

Dial-Up Account

This refers to having registered permission to access a remote computer by which you are allowed to connect through a modem.

Domain

One of the different subsets of the Internet. The suffix found on the host name of an Internet server defines its domain. For example, the host name for Prentice Hall, the publisher of this book, is www.prenhall.com. The last part, .COM, indicates that Prentice Hall is a part of the commercial domain. Other domains include .MIL for military, .EDU for education, .ORG for non-profit organizations, .GOV for government organizations, and many more.

Download

The process of transferring a file, document, or program from a remote computer to a local computer. (See Upload.)

E-mail

The short name for electronic mail. E-mail is sent electronically from one person to another. Some companies have e-mail systems that are not part of the Internet. E-mail can be sent to one person or to many different people.

Encryption

A security procedure of coding information to prevent unwanted viewing. Information sent across a computer network is typically disassembled, shipped, and reassembled on the receiving computer. Encrypted information must be decrypted with a special "encryption key" by the receiving party.

Executable File

A file or program that can run (execute) by itself and that does not require another program. Some files, such as word processor documents, require an applications program for viewing them.

External Viewer Application

Browsers are software applications that enable users to display content distributed on the Web. Web information must be in one of a few specific formats before the browser can display it for the users. An External Viewer Application can be used to view files sent across the Web that cannot be viewed within the browser. These applications are said to be external because they do not operate within the browser. (See Plugin.)

FAQ

This stands for frequently asked questions. A FAQ is a file or document where a moderator or administrator will post commonly asked questions and their answers. Although it is very easy to communicate across the Internet, if you have a question, you should check for the answer in a FAQ first.

Firewall

A firewall is a network server that functions to control traffic flow between two separate networks. They are typically used to separate large government and corporate sites from the Internet. Some colleges use firewalls to protect certain areas of their network.

Flame

Degrading a person over the Internet is referred to as flaming. Non-verbal communication is not typically possible across a computer network, unless you have a video hookup, so misunderstandings often result. Anonymity of the flamer also contributes to such an exchange because people are more likely to make impolite statements given their physical separation.

FTP

This stands for File Transfer Protocol. It is a procedure used to transfer large files and programs from one computer to another. Access to the computer to transfer files may or may not require a password. Some FTP servers are set up to allow public access by anonymous log-on. This process is referred to as Anonymous FTP.

GIF

This stands for Graphics Interchange Format. It is a format created by CompuServe to allow electronic transfer of digital images. GIF files are a commonly used format and can be viewed by both Mac and Windows users.

Gopher

A format structure and resource for providing information on the Internet. It was created at the University of Minnesota.

GUI

An acronym for Graphical User Interface. Macintosh and Windows operating systems are examples of typical GUIs.

Helper

This is software that is used to help a browser view information formats that it couldn't normally view.

Hits

This refers to a download request made by a browser to a server. Each file from a Web site that is requested by the browser is referred to as a hit. A Web page may be composed of numerous file elements and although hit counts are often reported as a measure of popularity, they can be misleading.

Home Page

In its specific sense, this refers to a Web document that a browser loads as its central navigational point to browse the Internet. It may also be used to refer to as Web page describing an individual. In the most general sense, it is used to refer to any Web document.

Host

Another name for a server computer. (See Server.)

Hotlist

This is a list of URLs saved within the Mosaic Web browser. This same list is referred to as a Bookmark within the Netscape Web browser.

HTML

This is an abbreviation for HyperText Markup Language, the common language used to write documents that appear on the World Wide Web.

HTTP

An abbreviation for HyperText Transport Protocol, the common protocol used to communicate between World Wide Web servers.

Hypertext

An embedded connection within a Web page that connects to a site within the viewed Web page or to a different Web page. Web pages use hypertext links to call up documents, images, sounds, and video files. The term hyperlink is a general term that applies to elements on Web pages other than text elements.

Icon

This refers to a visual representation of a file or program as it is represented on a typical windows graphic user interface (GUI). For example, Apple uses a trashcan icon to represent the place to put files you want to delete or remove from your computer. Microsoft uses a wastepaper basket.

Internet Relay Chat (IRC)

IRC is a network attached to the Internet. It allows users to converse in real time with other individuals. It is not typically a one-on-one conversation. Chat "rooms" are typically a very confusing place for beginners.

Internet Service Provider (ISP)

A company that provides Internet access is an ISP. Your ISP might be your school or a company to which you subscribe on a monthly basis.

Intranet

This refers to a network of networks that does not have a connection to THE Internet.

ISDN

This stands for Integrated Services Digital Network. It is a digital phone line. ISDN service is typically more expensive but also offers customers added features such as a greater bandwidth. (See Bandwidth.)

Java

An object-oriented programming language developed by Sun Microsystems.

JavaScript

A scripting language developed by Netscape in cooperation with Sun Microsystems to add functionality to the basic Web page. It is not as powerful as Java and works primarily from the client side.

JPEG

This stands for Joint Photographic Experts Group. It is one of the common standards for pictures on the Internet.

Local Area Network (LAN)

A LAN is a small or local network, typically within a single building.

Link

A text element or graphic within a document that has an embedded connection to another item. Web pages use links to access documents, images, sounds, and video files from the Internet, other documents on the local Web server, or other content on the Web page. Hyperlink is another name for link.

List Administrator

An individual that monitors or oversees a mailing list. (See Mailing List.)

Login

Generally, this refers to the act of connecting to a network but it may also indicate the need to enter a username or password to access a network or server.

Lurker

An individual who connects to a chat room, bulletin board, or newsgroup and observes the conversation but does not participate.

Mailing List

A functional group of e-mail addresses intended for making group mailings. It is used as a simple bulletin board. Some mailing lists are moderated by an individual and some are automatic. The most common mistake made by people using mailing lists is that they reply to a message and forget that everyone on the list will receive and potentially read their note. This can have embarrassing consequences.

MIME Type

This stands for Multipurpose Internet Mail Extension. It is a standard used to identify files by their extension or suffix. Applications, like your e-mail client, are said to be MIME compliant when they can decode MIME suffixes. (See MOV, MPG, PDF.)

Mirror site

Some sites on the Internet are very popular and under heavy demand by the viewing public and are potentially overloaded with traffic. Mirror sites are exact copies of the original site that help to distribute the traffic load, increasing efficiencies in delivering information.

Modem

A modem is a device used to send and receive information across a phone line by your computer. Computers speak digital and telephones speak analog. Essentially, a modem is a translator. Modems are only one kind of device available for connecting your computer to the outside world. Two other methods becoming more common for home use are ISDN and cable.

MOV

This stands for movie. It is a file extension for animations and videos in the QuickTime file format.

MPG/MPEG

This stands for Motion Picture Experts Group. It is a format for both digital audio and digital video files.

Multimedia

As a general definition, multimedia is the presentation of information by multiple media formats, such as words, images, and sound. Today, it's more commonly used to refer to presentations that use a lot of computer technology.

Nettiquette

This is a word created to mean Network Etiquette. It is a general list of practices and suggestions to help preserve the peace on the Internet. (See Flame.)

Newsgroup

This is the name for the discussion groups that can be on the *Usenet*. Not all newsgroups are accessible through the Internet. Some are accessible only through a modem connection. (See *Usenet*.)

Pathname

A convention for describing or outlining the location of a file or directory on a host computer. A URL is typically composed of several elements in addition to the pathname. For example, in this URL: http://www.prenhall.com/pubguide/index.html, http:// describes the protocol for a Web server, www.prenhall.com is the name of the host or server, /pubguide/ is the pathname, and index.html is the file name.

PDF

This stands for Portable Document Format. It is a file format that allows authors to distribute formatted, high-resolution documents across the Internet. A free viewer, Adobe Acrobat Reader, is required to view PDF documents.

Plugin

This is a resource or program that can be added to a browser to extend it function and capabilities.

QuickTime (QT)

A file format developed by Apple Computer so that computers can play digital audio, animation, and video files. (See MOV, MPG.)

Robot

An automated program used to search and explore the Internet. Some popular search engines use these programs.

Search Engine

An on-line service or utility that enables users to query and search the Internet for user-defined information. They are typically free services to the user. (See Robot.)

Search String

A logical collection of terms or phrases used to describe a search request. Some search engines enable the user to define strings with Boolean cues such as AND, NOT, or OR. (See Search Engine.)

Server

A software program used to provide, or serve, information to remote computers. Servers function in a Client-Server information exchange model. This term may also be loosely applied to the computer that is used to serve the information. (See Client.)

Shareware

Software that is provided to the public on a try-before-you-buy basis. Shareware functions on the honor system. Once you've used it for a while, you are expected to pay a small fee. Two similar varieties of software are Freeware and Postcardware. Freeware is just that, free for your use and the owners of Postcardware simply ask you to send them a postcard to thank them for the product.

Shockwave

Shockwave is a plugin that allows Macromedia programs to be played on your Web browser. Many learning tools are beginning to be posted to the Web as Shockwave files. Visit Macromedia's Web site for more information and to download the plugin (http://www.macromedia.com/). (See Plugin.)

Signature

A signature is text that is automatically added to the bottom of electronic communications such as e-mail or newsgroup postings. A signature usually lists the name and general information about the person making the posting. Using a signature means that you don't have to repeatedly type your name and return information every time you send a note.

SPAM

The electronic version of junk mail. It also refers to the behavior of sending or posting a single note to numerous e-mail or newsgroup accounts. It is considered to be very bad nettiquette.

Stuff

The action of compressing a file using the Stuff-It program. This is a Macintosh format.

Table

A specific formatting element found in HTML pages. Tables are used on HTML documents to visually organize information.

Telnet

The process of remotely connecting and using a computer at a distant location.

Thread

This describes a linked series of newsgroup postings. It represents a conversation stream. Messages posted on active newsgroups are likely to spur numerous replies each of which can spin off into an independent conversation. The nature of newsgroups allows a reader to move forward or backward through a conversation as if moving along a string or thread.

Topic Drift

This describes the phenomena observed in many on-line conversations, typically chat or newsgroup, where the topic will drift or change from the original posting.

Upload

The process of moving or transferring a document, file, or program from one computer to another computer.

URL

An abbreviation for Universal Resource Locator. In its basic sense, it is an address used by people on the Internet to locate documents. URLs have a common format that describes the protocol for information transfer, the host computer address, the path to the desired file, and the name of the file requested.

Usenet

A worldwide system of discussion groups, also called newsgroups. There are many thousands of newsgroups, but only some of these are accessible from the Internet.

User Name

An ID used as identification on a computer or network. It is a string of alphanumeric characters that may or may not have any resemblance to a user's real name.

Viewer

A program used to view data files within or outside a browser. (See External Viewer Application.)

Virtual Reality (VR)

A simulation of three-dimensional space on the computer. (See VRML.)

VRML

This stands for Virtual Reality Markup Language. It was developed to allow the creation of virtual reality worlds. Your browser may need a specific plug-in to view VRML pages.

WAV

This stands for Waveform sound format. It is a Microsoft Corporation format for encoding sound files.

Web (WWW)

This stands for the World Wide Web. When loosely applied, this term refers to the Internet and all of its associated incarnations, including Gopher, FTP, HTTP, and others. More specifically, this term refers to a subset of the servers on the Internet that use HTTP to transfer hyperlinked document in a page-like format.

Webmaster

This is the general title given to the administrator of a Web server.

Web Page

A single file as viewed within a Web browser. Several Web pages linked together represent a Web site.